AF270244

Animals in Streams & Rivers

Julie Murray

Abdo Kids Junior
is an Imprint of Abdo Kids
abdobooks.com

Abdo
ANIMAL HABITATS
Kids

abdobooks.com

Published by Abdo Kids, a division of ABDO, P.O. Box 398166, Minneapolis, Minnesota 55439.
Copyright © 2021 by Abdo Consulting Group, Inc. International copyrights reserved in all countries.
No part of this book may be reproduced in any form without written permission from the publisher.
Abdo Kids Junior™ is a trademark and logo of Abdo Kids.

Printed in China

052020

092020

THIS BOOK CONTAINS
RECYCLED MATERIALS

Photo Credits: Alamy, iStock, Shutterstock

Production Contributors: Teddy Borth, Jennie Forsberg, Grace Hansen

Design Contributors: Candice Keimig, Pakou Moua, Dorothy Toth

Library of Congress Control Number: 2019955550

Publisher's Cataloging-in-Publication Data

Names: Murray, Julie, author.

Title: Animals in streams & rivers / by Julie Murray

Description: Minneapolis, Minnesota : Abdo Kids, 2021 | Series: Animal habitats | Includes online resources
and index.

Identifiers: ISBN 9781098202118 (lib. bdg.) | ISBN 9781098203092 (ebook) | ISBN 9781098203580
(Read-to-Me ebook)

Subjects: LCSH: Animals--Habitations--Juvenile literature. | Habitat (Ecology)--Juvenile literature. |
Stream animals--Juvenile literature. | Rivers--Juvenile literature. | Stream ecology--Juvenile literature.

Classification: DDC 591.52--dc23

Table of Contents

Animals in Streams & Rivers

Streams and rivers have moving water. Many animals live in them.

Turtles sit on rocks.

They soak up the sun.

Alligators are big. Some can
be 15 feet (4.5 m) long!

Catfish have **barbels**. They

look like whiskers.

barbel

Frogs have long tongues.

They use them to catch bugs.

13

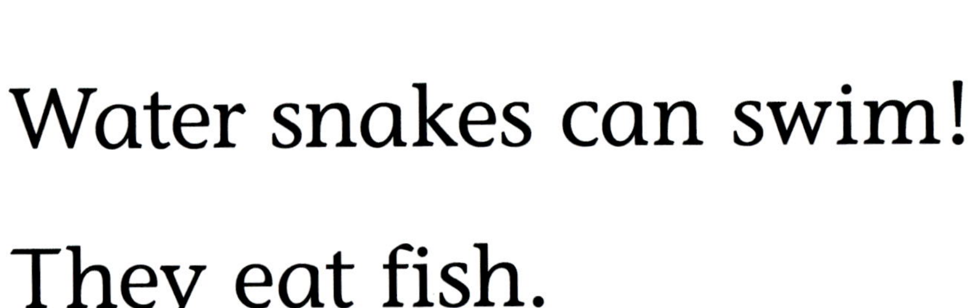

Water snakes can swim!

They eat fish.

Snails have hard shells.

Shells **protect** them.

Trout swim in the water.

Some are very colorful.

River otters have **webbed** feet.

These help them swim.

More Animals in Streams & Rivers

bluegill

crayfish

smallmouth bass

water beetle

Glossary

barbel
a whisker-like feeler made of skin that helps a fish sense food.

protect
to keep safe from harm.

webbed
having toes connected by thin skin.

Index

Visit **abdokids.com** to access crafts, games, videos, and more!

Use Abdo Kids code
AAK2118
or scan this QR code!